CIRIA C657  London, 2006

Books are to be returned on or before the last date below.

DUE CHECK ON RETURN FOR CD/DISK

0 5 OCT 2010

ks for

CIRIA *sharing knowledge* ■ *building best practice*

Classic House, 174–180 Old Street, London EC1V 9BP
TELEPHONE 020 7549 3300   FAX 020 7253 0523
EMAIL enquiries@ciria.org
WEBSITE www.ciria.org

Summary

With the increasing requirement to safeguard water supplies in the UK and implement sustainable water use, managers of offices and hotels need to know what the indicators are of national trends and what benchmarks are appropriate for their buildings.

This report describes a study that has used data from a number of water companies to establish the ranges of usage at present and to set benchmarks for better practice. It will enable managers to understand whether their office or hotel is performing well or poorly in comparison with other similar buildings in the UK. Guidance is provided on actions that can be taken to reduce usage. It will also inform designers about target usage and will be of value to those who wish to carry out similar benchmarking studies and future researchers.

Water key performance indicators and benchmarks for offices and hotels

Waggett, R; Arotsky, C

CIRIA

Publication C657   © CIRIA 2006   RP647   ISBN   0-86017-657-6
978-0-86017-657-2

British Library Cataloguing in Publication Data

A catalogue record is available for this book from the British Library.

Keywords		
Water resources, sustainable resource use, water infrastructure, facilities management, internal environment, environmental good practice, sustainable construction, building technology		
Reader interest	**Classification**	
Facility managers, building owners and developers, employers, water engineers and advisors, water undertakers, regulatory bodies, surveyors, architects, designers	AVAILABILITY	Unrestricted
	CONTENT	Advice/guidance
	STATUS	Committee-guided
	USER	Asset holders, water undertakers, facility managers, consultants, regulators

Published by CIRIA, Classic House, 174–180 Old Street, London EC1V 9BP, UK.

All rights reserved. No part of this publication may be reproduced or transmitted in any form or by any means, including photocopying and recording, without the written permission of the copyright-holder, application for which should be addressed to the publisher. Such written permission must also be obtained before any part of this publication is stored in a retrieval system of any nature.

This publication is designed to provide accurate and authoritative information in regard to the subject matter covered. It is sold and/or distributed with the understanding that neither the author(s) nor the publisher is thereby engaged in rendering a specific legal or any other professional service. While every effort has been made to ensure the accuracy and completeness of the publication, no warranty or fitness is provided or implied, and the author(s) and publisher shall have neither liability nor responsibility to any person or entity with respect to any loss or damage arising from its use.

Acknowledgements

Research contractor	Faber Maunsell
Lead author	**Rachel Waggett BSc (Hons)**

Rachel is an associate director in Faber Maunsell's Sustainable Development Group. She is a sustainability advisor with more than ten years experience specialising in providing water conservation advice for commercial and residential buildings. Previous publications on subjects including water conservation, greywater and rainwater reuse are augmented by regular participation in the design of water facilities in new and refurbished buildings of all types.

Co-author **Catherine Arotsky, BEng (Hons), MSc, CEng, MEI**

Catherine is a senior consultant at Rickaby Thompson Associates. She has over seven years experience as a sustainability adviser advising on energy and environment in buildings and has particular expertise in monitoring buildings and analysing data. Catherine is a member of the Energy Institute, and has worked extensively with housing associations and local authorities, undertaking specification reviews, providing energy efficiency and sustainability advice, and monitoring innovative technologies.

Steering group Following CIRIA's usual practice, the research project was guided by a Steering Group which comprised:

Chair	Martin Osborne	Ewan Group (formerly Earth Tech Engineering)
Members	Darren Cook	Hilton Hotels Group
	David Calderbank	Environment Agency
	Alan Daw (*previously Sue Craddock*)	Thames Water
	Nigel Foster	Bournemouth and West Hampshire Water
	Meyrick Gough (*previously Magda Styles*)	Southern Water
	Gordon Hall	Dwr Cymru
	Adrian Jackson-Robbins (*previously Lawrence Mbugwa*)	Davis Langdon Consultancy (for DTI)
	Ian Newberry	Three Valleys Water
	Clare Ridgewell (*previously Sarah Bowerman*)	Essex and Suffolk Water
	Sue Roaf	Oxford Brookes University
	Martin Shouler	Arup (formerly BRE)
	Simon Walster (*previously Dene Bridge and Nicola Simpson*)	Ofwat
	Jon Wood	South West Water
	Alastair Wright	Scottish Water Solutions (formerly East of Scotland Water)

CIRIA's research manager for this project was Paul Shaffer.

Project funders

The project was funded by:

Bournemouth and West Hampshire Water
DTI Partners in Innovation
Scottish Water (formerly East of Scotland Water)
Environment Agency
Essex and Suffolk Water
Ofwat
Southern Water
South West Water
Vivendi
Welsh Water

Executive summary

Increasingly, there is a requirement to safeguard water supplies in the UK and implement sustainable water use. Practitioners working to advise the office and hotel sectors on reductions in water consumption had identified a lack of suitable benchmarking information that would allow a simple method of comparing the water consumption of a particular property with national trends.

This report describes how suitable benchmarks have been found for water use in offices and hotels across England and Wales that will enable managers to understand whether their property is performing well or poorly in comparison with other similar buildings in the UK. It will also inform designers about target usage.

This detailed report should be read by those who are interested in how the research was carried out, or who wish to carry out similar benchmarking studies. It also includes details of the information gained by the research team that may be of use to future researchers.

Guidance reports

For those who do not require the detail in this report, the benchmarks and some guidance to assist improvement in water consumption have been provided in separate guidance reports freely available to download from the internet:

W10 – *Key performance indicators for water use in hotels*
www.ciria.org/downloads/01/w010.pdf

W11 – *Key performance indicators for water use in offices*
www.ciria.org/downloads/01/w011.pdf

CIRIA encourages wide dissemination of these guidance reports.

The key performance indicator chosen for hotels is **cubic metres of water per bedspace per annum**. Although the occupancy rate for the hotel is the main defining factor for water consumption, it was found that few small and medium sized hotels (the main target of this research) have the facility to easily collect guest night information. Subsequently, bedspaces were used as a proxy.

The analysis showed a major correlation between the "star" rating of the hotel and the water consumption. However, "star" rating is a voluntary exercise and many hotels choose not to have "star" ratings. In this case they have been grouped as "other" hotels.

Key performance indicators for offices have been provided both for water consumption in **cubic metres per person per year**, and **cubic metres per square metre per year**. Occupancy was again thought to be the main determinant of water consumption, but it was recognised that occupancy can be a difficult and ambiguous figure to find for a typical office building. The alternative was to define a KPI based on floor area, recognising that this is not as accurate a measure, but one that can be readily determined.

Suitable benchmarks have been compiled from this research project for office and hotel buildings across the UK. However the available data was limited and CIRIA and the sponsoring water companies hope that by making the data available with this report (see the attached CD-Rom at the back of the book), others will be able to build on the current research and refine the proposed benchmarks.

CONTENTS

Summary 2
Acknowledgements 3
Executive summary 5
List of figures 7
List of tables 8
Glossary 9

1   Introduction and background 10
    1.1   The project need 10
    1.2   Who should read this report? 11
    1.3   Guidance report 11
    1.4   Definitions of benchmarking and key performance indicators . . . 11

2   Summary of key points 13
    2.1   Summary of proposed benchmarks and key performance
         indicators 13
        2.1.1   Summary of hotels sector 13
        2.1.2   Summary of offices sector 14
    2.2   Further work/lessons learned 15
        2.2.1   Further work 15
        2.2.2   Lessons learned 16

3   Methodology 18
    3.1   Developing the key performance indicators 19
    3.2   Obtaining data 20
    3.3   Data manipulation and cleaning 21
    3.4   Benchmarks 21
    3.5   Analysis 21
    3.6   Lessons learned 22
        3.6.1   Obtaining and cleaning water company data 22
        3.6.2   Correlation between benchmarks 24

4   Benchmarking hotels 25
    4.1   Water use in hotels 25
        4.1.1   Thames Water 25
        4.1.2   Welsh hotels water demand management 26
        4.1.3   International Hotels Environment Initiative (IHEI) 26
    4.2   Defining a key performance indicator 26
    4.3   Other water consumption variables 28
    4.4   Obtaining data 29
        4.4.1   Separating data into water company areas 29
    4.5   Hotel data analysis 30
        4.5.1   Initial correlation and aggregated analysis 31

		4.5.2 Dividing hotels into types for further analysis 32
	4.6	Hotel water consumption by star rating 34
	4.7	Hotel water consumption by star rating and presence of swimming pool 35
	4.8	Hotels with swimming pools by category 38
	4.9	Recommended hotel benchmarks 42
5	**Benchmarking offices** **46**	
	5.1	Water use in offices 46
		5.1.1 Thames Water 46
		5.1.2 Government estate 47
		5.1.3 Watermark 47
	5.2	Defining a key performance indicator 47
	5.3	Other water consumption variables 48
	5.4	Obtaining data 49
	5.5	Office data analysis (floor area) 50
	5.6	Office data analysis (occupancy) 51
		5.6.1 The method of calculating personnel numbers 53
		5.6.2 The type of building used in the research 54
	5.7	Recommended office benchmarks 55
7	**References** **56**	

Figures

Figure 3.1	Project methodology 18
Figure 4.1	Chart showing frequency distribution of water consumption per bedspace for consumption in 2002 32
Figure 4.2	Chart showing frequency distribution of water consumption for hotels with swimming pools for consumption in 2002 33
Figure 4.3	Chart showing the frequency distribution of water consumption per bedspace for those hotels without swimming pools for 2002 33
Figure 4.4	Water consumption of hotels up to three star rating 34
Figure 4.5	Water consumption of four and five star hotels 34
Figure 4.6	Chart showing frequency distribution of water consumption per bedspace for "other" category hotels 35
Figure 4.7	Frequency distribution of water consumption (m^3) per bedspace for category 1 hotels 36
Figure 4.8	Frequency distribution of water consumption (m^3) per bedspace for category 2 hotels 37
Figure 4.9	Frequency distribution of water consumption (m^3) per bedspace for category 3 hotels. 38
Figure 4.10	Frequency distribution of water consumption (m^3) per bedspace of hotels with no stars and with swimming pools. 39
Figure 4.11	Chart showing frequency distribution for water consumption per bedspace for hotels with no star and with no swimming pool for consumption in 2002 39
Figure 4.12	Chart showing frequency distribution for water consumption per bedspace for hotels with one star and no swimming pool for consumption in 2002 . . . 40

Figure 4.13	Chart showing the frequency distribution of water consumption in hotels with two or three stars with swimming pool for consumption in 2002	40
Figure 4.14	Chart showing frequency distribution of water consumption per bedspace for hotels with two or three stars without a swimming pool for consumption in 2002	41
Figure 4.15	Chart showing frequency distribution of water consumption per bedspace for hotels with four or five stars and a swimming pool for consumption in 2002	41
Figure 4.16	Chart showing frequency distribution of water consumption per bedspace for hotels with four or five stars and no swimming pool for consumption in 2002	42
Figure 4.17	Chart showing benchmarks for each category of hotel without pools	43
Figure 4.18	Chart showing benchmarks for each category of hotel with pools	43
Figure 4.19	Chart showing CIRIA benchmarks for hotels with pools against Thames Water study benchmarks.	44
Figure 5.1	Chart showing frequency distribution for consumption by floor area	51
Figure 5.2	Chart showing frequency distribution for consumption per total employee.	52

Tables

Table 2.1	Benchmarks for hotels with swimming pools	13
Table 2.2	Benchmarks for hotels without swimming pools	14
Table 2.3	Benchmarks for offices	14
Table 3.1	Preferred sample size for analysis	20
Table 3.2	Standards of water usage for benchmarking	21
Table 4.1	Thames Water hotels benchmarks	25
Table 4.2	Water companies and hotels allocated to each	30
Table 4.3	Hotel sample sizes	31
Table 4.4	Overall hotel water usage	32
Table 4.5	Benchmarking outputs for hotels	42
Table 4.6	Recommended benchmarks for hotels with swimming pools	44
Table 4.7	Recommended benchmarks for hotels without swimming pools	45
Table 5.1	Thames Water study outputs for offices	46
Table 5.2	Water companies and offices allocated to each	50
Table 5.3	Recommended benchmarks for offices by floor area	51
Table 5.4	Recommended benchmarks for offices by employee numbers	53
Table 5.5	Table showing comparison of proposed benchmark and other research benchmarks	53
Table 5.6	Recommended combined office benchmarks	55

Glossary

Benchmark	The "best in class" level of performance achieved for a specific business process or activity. It is used as a reference for comparison in benchmarking.
Co-efficient of determination	See R-squared value below.
Correlation co-efficient	See R-squared value below.
Histogram analysis tool	This analysis tool calculates individual and cumulative frequencies for a range of data and data bins. It generates data for the number of occurrences of a value in a data set. A histogram table presents the chosen boundaries and the number of scores between the lowest bound and the current bound. The single most frequent score is the mode of the data.
Key performance indicator (KPI)	The measure of performance associated with an organisation's activity or process. This information provided by a KPI can be used to determine how an organisation compares with the benchmark and is therefore a key component in an organisations move towards best practice.
R-squared value	Also known as the **coefficient of determination** or the **correlation coefficient.** An indicator that ranges in value from 0 to 1 and reveals how closely the estimated values for the trendline correspond to actual data. A trendline is most reliable when its R-squared value is at or near 1.
Regression line	See trendline.
Trendline	A graphical representation of the trend or direction of data in a series. Most valuable when at or near to 1.
Net Internal Area (NIA)	A common figure used to represent commercial space that excludes core and plant (non-lettable) space. Also known as Net Lettable Area (NLA).

1 Introduction and background

This report summarises an approach to find suitable benchmarks for water use in offices and hotels across the UK. The aim is to enable managers of offices and hotels to understand whether their property is performing well or poorly in comparison with other similar buildings in the UK, and to inform designers about target usage.

1.1 The project need

The project need was identified from the increasing requirement to safeguard water resources and supplies in the UK and implement sustainable water use. Practitioners working to advise the office and hotel sectors on reductions to water consumption had identified a shortage of useful data. In particular there was a lack of suitable benchmarking information facilitating a simple method of comparing the water consumption of a particular property with national trends.

Similar information is available for energy use in offices and other buildings, and is regularly used at both the design and operation/management phases of buildings to identify whether action is required. The aim of this project was to provide similar data for water consumption which is accessible for use by interested parties including property managers, owners and operators.

The objectives of this project were to:

- collate and analyse existing data and information that is representative of UK water use
- provide facility managers, building engineers and designers with key performance indicators (KPIs) from which benchmarks for water efficiency in domestic, commercial and hotel environments can be set
- help improve water efficiency and encourage the sustainable use of water within buildings
- set up and manage a network of stakeholders – including developers, engineers, regulators, water utilities, government and consumers – to share information and disseminate guidance to end users
- make recommendations for further work.

These original objectives included the production of a KPI for commercial buildings which would include factory buildings. However this was modified following discussion with the project steering group at an early stage, given that factory process use is difficult to predict and already adequately covered by Envirowise guidance information[1]. Non-process use would be essentially office based and could be dealt with by the provision of an office benchmark from this project.

It was also originally intended to include residential use within the study, but the availability of data for the residential element proved extremely problematic and it was agreed that the study should analyse just the hotels and offices sectors. These sectors are described in detail in this report.

1.2 Who should read this report?

This project report provides a detailed summary of the project methodology and results, and should be read by those who want to understand how the research was carried out, or wish to carry out similar benchmarking studies. This document also includes details of the lessons learned by the research team that it is felt will be of considerable use for future research.

The questions raised by the study will not only be of interest to those involved in benchmarking for water consumption, but many of the lessons learned are also applicable to energy benchmarking, or those involved in any type of benchmarking in the offices and hotels sectors. It is hoped that by considering these issues while planning future benchmarking studies, many of the difficulties experienced during this project can be overcome by future researchers.

This report contains the following principal sections:

Section 2: Summary of key points; summarises the main points of the study and the findings, including lessons learned.

Section 3: Methodology; outlines the method followed by the research team and the reasons for the methodology chosen, where appropriate. Describes the process of KPI development and how data was obtained and analysed. A more detailed discussion of lessons learned is also included.

Section 4: Benchmarking hotels; describes the approach taken to produce information on water consumption in hotels including details of published and unpublished studies that helped to inform the research team. Proposes benchmarks for hotels and outlines the suggested split of hotels into typologies to assist benchmarking.

Section 5: Benchmarking offices; describes the approach taken to produce benchmarks for offices including a review of benchmarks currently utilised in the offices sector. The process of obtaining and analysing data is described, and proposed benchmarks explained.

1.3 Guidance reports

For those who do not require the detail in this report, the benchmarks and some guidance to assist improvement in water consumption have also been provided as separate guidance for hotels and for offices. It is intended that the content of the guidance reports can be used by third parties within specifications, guidance documents and information leaflets, for example, and therefore wide dissemination is encouraged.

Further guidance can be found in the CIRIA publications *Rainwater and greywater use in buildings. Decision-making for water conservation* (D Leggett *et al*, 2001), *Rainwater and greywater use in buildings. Best practice guidance* (D Leggett *et al*, 2001) and *Sustainable water management in land use planning* (P Samuels *et al*, 2005).

The guidance reports can be downloaded from the internet at:
www.ciria.org/downloads/01/w010.pdf and **www.ciria.org/downloads/01/w011.pdf**

1.4 Definitions of benchmarking and key performance indicators

Benchmarking involves comparing and measuring performance against others in key business activities and then using lessons learned from the best to make targeted improvements. It requires answering two questions – who is better and why are they better? With the aim of using this information to make changes that will lead to vital improvements.

Definitions of terms used within this report are contained in the glossary, but to aid understanding it is useful to define these two important terms at this point:

- a **key performance indicator** (KPI) is the measure of performance associated with an activity or process. This information provided by a KPI can be used to determine how an organisation compares with the benchmark and is a key component in a move towards good practice. An example of a KPI used in this report is water consumption per employee per annum.
- a **benchmark** is a level of performance achieved for a specific business process or activity. It is used as a reference for comparison in benchmarking.

The distinction between key performance indicators and benchmarks is that the KPI is the actual measure of performance whereas the benchmark is a target performance.

A CD-Rom accompanies this report which contains two excel files; one for data used in hotel analysis and the other for data in offices. All identifying information has been removed from this data following data protection requirements. However the structure of the data will enable it to be used for further analysis.

The remainder of this report discusses the results of the analyses currently performed.

2 Summary of key points

2.1 Summary of proposed benchmarks and key performance indicators

2.1.1 Summary of hotels sector

The key performance indicator for hotels was chosen as **cubic metres of water per bedspace per annum**. Previous studies described below found that occupancy of the hotel is the main defining factor for water consumption. It was determined that few small and medium sized hotels (the main target of this research) have the facility to easily collect guest night information (which would provide a more accurate KPI figure) therefore bedspaces were used as a proxy. More information on the selection of this benchmark is provided in Section 4.2. Cubic metres of water were used to enable easy comparison with bills.

The analysis determined that there was a significant correlation between the "star" rating of the hotel and water consumption. A concern with this system is that the star rating is a voluntary exercise and therefore many hotels do not have "star" ratings. In this case they have been grouped as "other" hotels, or those which are "undefined". The star ratings are also awarded for hotel facilities that will not affect water use, such as disabled facilities. Nevertheless a clear correlation was found and the benchmarks were defined for each of the four categories of hotel identified, as follows:

Category 1: 1 star rated establishments.
Category 2: 2 or 3 star rated establishments.
Category 3: 4 or 5 star rated establishments.
Other: Undefined establishments.

Further analysis determined that there was still at least one major factor unidentified, and the analysis indicated that this could largely be accounted for by sorting the data by the availability of a swimming pool at the hotel.

Benchmarks have been split into two distinct types: those with a pool and those without. They have further been split into the respective category of hotel. The benchmark figures shown below have also been rounded to provide ease of use.

Table 2.1   Benchmarks for hotels with swimming pools

Category	Hotel rating	Benchmarks (m³/bedspace/annum)		
		Best practice	Typical	Above average
Cat 1*	1 star	9	25	60
Cat 2	2 or 3 star	20	60	185
Cat 3	4 or 5 star	60	130	220
Other	No rating	40	90	170

* These figures have been derived as there were no category 1 hotels with pools

Table 2.2 Benchmarks for hotels without swimming pools

Category	Hotel rating	Benchmarks (m³/bedspace/annum)		
		Best practice	Typical	Above average
Cat 1	1 star	5	10	15
Cat 2	2 or 3 star	10	20	50
Cat 3	4 or 5 star	15	30	65
Other	No rating	10	30	70

2.1.2 Summary of offices sector

The KPI for the offices sector was difficult to define. There was a conflict between two of the essential components of a KPI; namely the need for the information to define it being easily available and unambiguous, and the need to reflect the main defining factor of an office building. In this case, the main defining factor was thought to be occupancy, but for various reasons it was realised that this is a difficult and ambiguous figure to find for a typical office building. The alternative KPI was office floor area, but this is not as accurate a measure as occupancy. Further details of the pros and cons of each KPI are provided in Section 5.2.

It was decided that both indicators should be tested by analysis to see which was the more appropriate: **water consumption in m³ per person per year** or **water consumption in m³ per square metre per year**. There was a preference for benchmarks to be stated in addition, as litres per day. To calculate the number of litres per day, a business year of 253 days was assumed, excluding weekends and bank holidays.

Analysis determined that both of these indicators had a strong correlation with the water consumption of the office buildings analysed, and both were suitable indicators. Analysis showed that there were no discernable additional factors that would need to be taken into account, and a single benchmark could be provided. Benchmarks were provided both by area and by total number of employees as shown in the table below. Where these could be verified, they compare favourably with previous research studies.

It was suggested that in order to provide appropriate guidance to the offices sector, it would be most useful to provide the typical benchmark only, encouraging facility managers to aim for a lower figure than this. This benchmark is highlighted in the table below.

Table 2.3 Benchmarks for offices

		Cubic metres per year	Litres per day (assuming 253 days per business year)
Typical use	By employee*	4.0 m³/employee/annum	15.8 litres/employee/day
	By area	0.6 m³/m²/annum	2.4 litres/m²/day
Best practice use	By employee*	2.0 m³/employee/annum	7.9 litres/employee/day
	By area	0.4 m³/m²/annum	1.6 litres/m²/day
Excessive use	By employee*	7.0 m³/employee/annum	27.7 litres/employee/day
	By area	0.8 m³/m²/annum	3.2 litres/m²/day

* Total employee numbers should be used even if full time equivalent employee figures are available.

Note: Figures have been rounded to form more usable benchmark figures.

2.2 Further work/lessons learned

2.2.1 Further work

Suitable benchmarks have been compiled from this research project for office and hotel buildings across the UK.

The main challenge in undertaking the work was the ability to obtain sufficient, robust data sources for analysis. The data set obtained required considerable analysis and data manipulation in order to "clean" the data sufficiently for a simple spreadsheet based analysis to be conducted, in line with the project objectives. Future work would be valuable to expand this set of data, as a larger data set would inevitably add to the robust nature and possible lessons learnt from the project.

Expand data set and identify further defining factors to improve benchmarks

By making the data available on the accompanying CD-Rom (additionally archived by CIRIA), it is possible that other research will be able to use this data freely to expand and build on current knowledge, refining the proposed benchmarks in the future. In particular, a larger data set may enable future researchers to identify further defining factors for offices and hotels, such as size of office etc.

Widen the data set to provide regional indicators, if significant

Producing a larger data set could enable regionality to be investigated. After "cleaning" the current sample was not large enough to enable this to be investigated, but it is a question that is often posed, and would be a very useful additional piece of work.

Identify areas where most water is used in offices and hotels, to enable improved monitoring and management

In the hotels sector this was undertaken to some extent by the Welsh Hotels study[5] although this study concentrated on replacing sanitaryware and measuring the impact, rather than recording in which areas the existing water consumed was being used (including hotel bedrooms, public toilets, bars, kitchens, etc). There is also no published information of this type about that project, which is a loss to water research in general.

In offices, currently the only existing breakdown of component water use is reliant on figures presented in 1996[2]. While these are a useful guide, the change in office premises, and particularly the likelihood of major differences between newly constructed or refurbished offices and older premises, means that a revision may be due. This would not need to consider in detail the total water consumption of an office, but the proportion of water used in each office function. A study of this type would enable refinement of the benchmarks to suit the case of the particular office building and would render them much more useful.

Identify trends in sanitaryware specification

Trends in sanitaryware are changing rapidly, but there is little information available concerning the changes over time, other than anecdotal. This makes prediction of reasons for changes in water consumption benchmarks over time very difficult to assess. Anecdotal evidence would suggest a move toward higher water consuming items within hotels, such as power showers, but changes in office specifications toward lower water use such as aerating taps, low volume toilets etc. However these changes happen quickly and do not appear to follow defined patterns. For example, in the last few years there has been a trend toward minimising or removing urinals from new offices with a preference for toilets and/or unisex washrooms. This will have a considerable impact, but there is little or no published information on such trends.

2.2.2 Lessons learned

The main lesson learned during this study is the immense difficulty of obtaining robust data for an exercise of this type. There are two main difficulties:

- obtaining sufficient information on the building to enable the defining factors to be identified, and explain the findings of the analysis
- obtaining water consumption data for large sets of data.

These challenges were also experienced when undertaking previous research. Both David Bartholomew Associates, Thames Water research and the Welsh Hotels research identify these difficulties specifically, despite great efforts to circumvent any potential issues during this project.

Two separate types of information were obtained.

1. The research team obtained a database of "offices" with "occupancy" information. The database contained "number of employees", but the database owners were unable to provide a definition of "number of employees".
2. The second database obtained was based on sales and rental information on floor area.

For both sources of data, no information was available on whether the data was for the whole building or which areas of the building were covered by which water meter. Floor area data is now available from the Valuation Office database, which outlines the different uses within a building. This data could be utilised, but there is still no reliable source of occupancy information.

The main challenge to undertaking water consumption benchmarking is obtaining robust and comparable water consumption information. Water metering information is held by water companies, but the databases are not predominantly designed to extract data based on water consumption for particular areas. Many systems are primarily designed for billing purposing rather than recording water use. It is recommended that water industry groups and particularly water companies who hold data on water consumption, consider how this data might be made more accessible should future studies be proposed.

Where reliable building information and reliable water consumption information was obtained, there were difficulties in correlating address information to the areas covered by water meters. Water meters often cover entire buildings but can cover areas within buildings. The issue of obtaining reliable water consumption data is examined in more detail in Section 3.6.

Correlation between benchmarks

By investigating both occupancy and floor area benchmarks, it was often found that there is no straightforward correlation between the two figures. Changes in office usage in recent years and in the future will make this even more problematic. Previous research where occupancy figures have been derived from floor area figures may not be as robust as imagined at the time, although based on industry accepted figures.

Similarly it was found that there was little correlation or agreement between existing published benchmark figures with, for example, total employees, total full-time equivalent employees, and per person figures all quoted for offices. A standard method of deriving benchmarks would be extremely helpful to managers and to future researchers. Again, water industry groups can play a lead role in this respect.

3 Methodology

The methodology for this study was determined by the aims and objectives of the project. In order to fulfil these aims, it was decided that the study should be:

- desk based
- readily replicable in the future
- based upon non-specialist software to enable simple replication and manipulation in the future
- appropriate to the majority of non-technical users of the outputs.

The flow chart below shows the methodology of the project, with the intended methodology illustrated on the left and the additional tasks that were necessary in order for the project to be completed shown on the right. An explanation of the lessons learned can be found in Section 3.6.

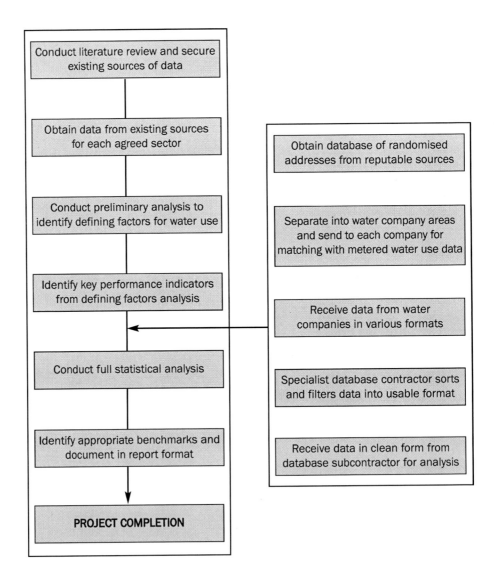

Figure 3.1   Project methodology

Initially it was intended that the study utilise existing data sets and simply analyse and expand these where necessary. However it quickly became clear that such data sets, either published or unpublished, did not exist, and that the study itself would need to obtain data for analysis. Details of this data and its origin is provided within each of the sector discussions on the following pages.

It was the intention of the project team to utilise available spreadsheet software throughout the study, to ensure that future researchers would be able to utilise the data produced by the study, and to repeat the methodology using standard software. However, the complexity of the data manipulation required the creation of a bespoke database in order to render the data usable for analysis and manipulation.

In order to meet the aim of the project, once the data had been cleaned of spurious or inaccurate records, all remaining data was transferred into excel (and appears in this format on the accompanying CD at the back of the book). All subsequent analysis was performed in excel and therefore if a future researcher has access to better source data it would be possible to repeat the analysis in a standard program.

3.1 Developing the key performance indicators

A key performance indicator (KPI) is the measure of performance associated with an activity or process. The information provided by a KPI can be used to determine how an organisation compares with others of similar type, and will directly influence a move toward good practice.

A KPI allows different performances to be directly compared with one another so that users can make informed choices and decisions. It is critical to the success of the KPI that it is:

- simple to use and understand
- sufficiently broad to be applied to as many buildings (in this case) as possible
- expressed in well defined standard units that allows easy conversion if necessary
- responds to the main factors that vary between buildings, such that the main difference in performance will be due to the influence of users or managers.

In defining KPIs for the sectors covered by this project, many lessons were drawn from KPIs used for benchmarking energy in buildings, a process that has a much longer track record than benchmarking of water use and has evolved during use to best suit market demands.

The KPI currently used for most energy benchmarking exercises is $kgC/m^2/year$ – kilograms of carbon per square metre per annum. This is a most useful and suitable indicator because:

- it is simple to use for all types of building, responding to the key defining factor for most buildings – their size (rather than occupancy which has a lesser effect on energy use in buildings)
- it responds to a secondary variation between buildings – their fuel source – by using carbon to measure performance rather than energy
- it is easy to convert other units into the benchmark unit (such as $kgCO_2/m^2$ or kgC expressed for the whole building)
- it uses a measure that is known for the majority of buildings – the floor area.

By utilising experience from the energy benchmarking processes, the key features for office use were considered. However, there are accepted limitations as well as the benefits of learning from the energy benchmarking process. Water benchmarks differ significantly from energy benchmarks, as the defining factors for energy (especially in the sectors considered) are much more evident. They tend to be responsive on a whole building level, as in general the buildings are heated and cooled as one entity, and are less responsive to user intervention. By comparison, water consumption is largely linked to user intervention and cannot be dealt with in the same manner.

Defining an appropriate KPI is a crucially important part of any benchmarking exercise and emphasis was placed on this phase of the project. The process of definition and conclusions for each sector are included in Sections 4.2 (for hotels) and 5.2 (for offices) respectively.

3.2 Obtaining data

Before obtaining data, the project team estimated the number of buildings within each sector in the UK. In order to ensure a statistically significant sample it was determined that a minimum of 0.5 per cent and preferably 1 per cent of the total number of buildings would be required. The suggested quantities are summarised below.

Table 3.1   Preferred sample size for analysis

	Total in UK	Minimum number of buildings	Preferred number of buildings
Offices[3]	278 000	1390	2780
Hotels[4]	60 000	300	600

These were minimum numbers in order to try to ensure statistical significance. The greater the number of records, the more robust the benchmark will be.

Having determined the number required, obtaining appropriate data proved to be the most challenging element of the project. As stated above, it was initially intended to obtain and analyse existing data sets, but it quickly became clear that no such data sets existed. New data had to be obtained for each sector. Data on office and hotel address and characteristics was obtained in both cases by purchasing a commercially available dataset. This had the benefit that a randomised sample could be obtained covering the whole of the UK. However there were disadvantages – most importantly that the quality of the data was very variable as it had been generated from responses to self completed questionnaires. This meant that a number of inconsistencies and inaccuracies were apparent in the data and these had to be investigated and removed manually.

The process of data collection was further complicated by the differences between water company procedures and reporting methodologies. While some water companies have sophisticated data collection databases that allow interrogation and export of large volumes of data in various formats, others had less sophisticated systems which were not able to match the information on hotels and offices to a water volume automatically. The process of completing the data set required for analysis became much more complicated, resource intensive and time consuming. Fortunately the majority of water companies who agreed to take part in the project were able to complete the matching of data, although in some cases this was done manually at great time and expense to the company concerned.

3.3 Data manipulation and cleaning

Due in part to the lack of automation in data collection, the data provided for analysis was unsuitable for direct use and a number of database routines were written to sort and order the data into a suitable format. Any data that could not be ordered was removed from the dataset and this led to a reduction in the amount of data available for analysis. The lessons learned from this process will be important for further studies and are described in more detail in Section 3.6 below.

In most cases, the metered water data collected represented a period either longer or shorter than a calendar year, a result of the differences in metering date. To ensure the analysis was accurate, each metered water reading was normalised to consider a single calendar year in each case.

3.4 Benchmarks

It was agreed with the project steering group that the following "splits" would be used in the formulation of benchmarks, as they concur with the majority of previous research and analysis conducted in the area of water benchmarking. A comparison with the benchmarks used by other research teams and previous publications is discussed in detail in the sectoral analysis of the hotel and offices sectors later in this document.

Table 3.2   Standards of water usage for benchmarking

Definition	Notes
Excessive use	The lowest quartile (75 per cent of buildings perform better than this target)
Typical practice	The median point of a set of data, or second quartile (50 per cent of buildings perform better than this target)
Best practice	The first quartile (25 per cent of buildings perform better than this target)

3.5 Analysis

One of the project aims was to ensure basic statistical analysis was undertaken to determine the distribution.

The initial stage was to determine whether there is a correlation within the data. There are a number of statistical methods for undertaking this, but for simplification, the inbuilt trendline function within spreadsheet software was utilised. A trendline was added together with the equation of the line and the R-squared (correlation) coefficient. Where the R-squared coefficient approaches a value of one (+1 or -1) this indicates that the data has a good correlation. The closer the value is to zero (0), the lower the correlation. The correlation indicates the degree to which two factors influence one another.

Once it was determined that the data had a reasonable correlation, the frequency distribution was calculated using the software's Histogram function. The results were plotted on a graph and the frequency interval adjusted depending on the results, to obtain a graph of the spread of results. From this, the three benchmarks above (25, 50 and 75 per cent) could be obtained. Further information on the analysis and how outliers were removed can be found in the relevant sectoral analysis.

The data was also analysed utilising the median, which is another way of presenting the data. This analysis showed that the use of percentage levels produced more appropriate results that were replicable. The use of percentage levels indicates an analytical approach which would provide more scope for replication in future projects. It also has the benefit of having been used by other research projects in the past, and produces directly comparable results.

3.6 Lessons learned

Obtaining sufficient, robust data to carry out a meaningful exercise has been the main difficulty for this research study. There are two main difficulties:

- obtaining sufficient information on the building to enable the defining factors to be identified, and analysis findings to be explained in a meaningful manner
- obtaining water consumption data for large sets of data.

Both of these items are discussed in more detail in this section. During the study, with access to unpublished water company research studies and information, it became evident that other researchers have experienced identical problems with obtaining data.

By documenting the information from this research, it is the hoped that in future exercises sufficient time and procedures will be incorporated into the planning stages so that any difficulties encountered are planned and managed, resulting in a net benefit to all future researchers in this important area.

3.6.1 Obtaining and cleaning water company data

The main barrier to undertaking research in this area is obtaining suitable water consumption information.

Matching address information to water consumption. Information on water consumption data for addresses in the UK is held by water companies. However each water company has its own procedures and reporting methodologies and some may be more focused on procedures for billing rather than measuring water consumption. While some water companies have sophisticated data collection databases that allow interrogation and export of large volumes of data in various formats, some have slightly less complex systems. These simpler, less sophisticated systems were not designed to match address information to a water volume automatically, and therefore the process of completing the data set required for analysis became much more complicated, resource intensive and time consuming. In some cases, this was carried out manually – completely unfeasible for larger data sets.

It is recommended that water industry groups and particularly water companies who hold data on water consumption, consider how this data might be made more easily accessible should future studies be contemplated.

Insufficient information to investigate irregular readings. Where water consumption data was obtained, it often appeared irregular, but there was no means of investigating such readings other than on an individual basis, which would be impractical for large data sets. There are several possible reasons why a consumption figure may be irregular, including:

- water meters often cover entire buildings or can cover areas within buildings, but there is no method of indicating when this occurs
- meter readings are often infrequent and/or estimated, and this is not always noted

- leakage or other unusual use of water is not known unless many years data are investigated.

It would appear that there is no practical method of avoiding these problems, and any evident irregular readings should be removed from the data set before analysis. To reduce the problem of data loss, a larger initial set than anticipated should be allowed for to cater for this eventuality.

Coding of meter reading type within water company databases. Coding of meter readings (to indicate for example, where a reading is estimated) has produced several difficulties. Some databases do not provide an indication as to which readings are estimated, meaning that without this data the information is flawed and unlikely to be usable. Some databases record some of this information, along with coding for different types of readings (ie estimated, actual, customer reading). The number of codes, and the variation between water companies using different codes, make sorting and analysis of larger data sets very time consuming with many analysis routines required. In some cases the coding used has evidently been input manually and this changes between lower and upper case notation (ie "E" or "e" for estimated readings). This seems a very simple difference but when automatically sorting data it requires a separate routine to be written.

Meter roll over information. All meters have a "roll over" figure which is the number at which the meter rolls back to zero. For some meters this will be 00000, whereas for others it may be 99999. In some cases it may be less logical, and this appeared to be the only explanation for some of the readings recorded. The data analysis also showed a large number of readings that had not changed for some years, presumably because the meter has been replaced, removed or had stopped working. These readings should be removed or they would affect analysis results.

Installation of new meters. Some water companies record details of when meters are replaced. For those which do not, it would appear that either new meters have been installed or there is a non-logical rollover. This is not clear from the data and these records have had to be searched for and deleted; a time consuming process for larger data sets. An assumption is required where it is possible the anomalous reading is due to a logical rollover. This also had to be undertaken manually.

Multiple meters. A larger than expected number of commercial buildings have multiple meters, and these should be searched for and either merged or removed to avoid several readings for a single address. Leaving them in the data set will affect the analysis and lead to substantially smaller readings than expected.

When considering analysing multiple water company data there is little that can be done to eliminate these problems. Setting and agreeing a standard format for the provision of data may be of assistance, but these could prove impossible, for all companies involved in this study found it difficult to adhere to the proposed format due to the nature of their databases. It is recommended that for multiple company exercises a considerable amount of time and effort is programmed into the research timetable to allow for these difficulties, which are likely to be insuperable in the short term. Certainly it would be impossible to carry out this exercise using a simple analysis programme such as excel, and a specialist database operator capable of writing numerous data sorting routines (with the appropriate computer hardware to manage such large data sets) should be included in all projects of this type.

For single water company consumption investigations, it is be preferable to begin with a data collection exercise to understand what the database is capable of producing and in what format, then working forward to the type of result that can be obtained from this data.

3.6.2 Correlation between benchmarks

Research indicated that little effort has been made by previous researchers and those publishing guidance to use benchmarks that correlate with one another and augment the various data sets.

For offices in particular, total employees, total full-time equivalent employees, and per person figures are all used in published guidance. This report explains the crucial differences between the figures generated by each indicator, and the importance of knowing and explaining the differences. Often they are published without sufficient information to guide a non-technical person on the importance and difference of each type of benchmark.

It is likely that in some cases a benchmark originally shown in one format has been transposed during publication, or for reasons of space allocation in documents, or "plain English", the explanations have been omitted. Not only does this lead to confusion and dilution of the key message promoted, but it does not assist in further work. Every effort has been made in this report and in the guidance that accompanies it, to make the differences and the format used very clear.

In addition, a recommendation has been made within this report for a coherent and single benchmark for both employee and office area, and bedspace (for hotels) respectively. Promotion by all water companies and all future researchers of this single benchmark figure would enhance the use of benchmarks and would reduce confusion. It would also assist Government and other national bodies who wish to see water consumption gain the same potential for benchmarking and measurement as energy, targets for which are now included in the Building Regulations and other legislation. Without such a single agreement, this will not be possible for water consumption as it will never be as widely used.

4 Benchmarking hotels

4.1 Water use in hotels

There are approximately 60 000 hotels in the UK of varying sizes[4]. Initial research determined that large hotels were already benchmarking their performance using a tool known as BenchmarkHotel. Following discussions with the project steering group, it was determined that this project should concentrate on small, independently owned hotels. Often, this group of hotels do not have the resources to undertake large benchmarking projects and research from other sources determined that small hotels have the least knowledge of water conservation[5] but conversely may be able to make proportionally large savings from reducing their water consumption.

The approach for investigating this sector was based on previous work and was agreed with the project steering group. It was necessary to consider not only the information that would be of most use to the hotels sector within the target size selected, but also the information that it was possible to obtain within the scope of the study. Further discussion of the use of information and selection of benchmarks and KPIs is provided below.

4.1.1 Thames Water

A study for Thames Water in 1999 investigated water use in hotels and proposed benchmarks for a number of hotel types[5]. This desk based study used data produced from Thames Water's customer information system regarding hotels in the Thames Water region, matching this manually to information from readily available tourist information guidance. The KPI utilised was m^3 per bedspace per year. Despite various difficulties in obtaining robust data, a total of 597 hotels were analysed and the following benchmarks proposed:

Table 4.1 Thames Water hotels benchmarks

Benchmarks m^3/bedspace/yr	Bed and breakfast	2 and 3 star	4 and 5 star	Approx %ile level
Low	<30	<40	<80	bottom 25%
Average	30-40	40-60	80-100	25-50%
High	40-70	60-110	100-230	50-90%
Very high	>70	>110	>230	top 10%

In order to provide further value to the study, a number of audits of hotels were also carried out and some valuable lessons highlighted as a result. The study also went on to estimate potential savings for hotel groups in the entire Thames region should this sector be encouraged to improve their water performance.

4.1.2 Welsh hotels water demand management

This study of Welsh Hotels[6] aimed to investigate the effect on hotel water use by installing various types of water efficient equipment, and monitoring the hotel in detail both before and after installation. In addition the hotels were compared against benchmarks created by analysis of water consumption data from many more hotels, in an analysis exercise similar to this research project.

A summary of the work has provided some information on the conclusions. In the study, a series of benchmarks was compiled for two and three star hotels utilising the Welsh Tourist Board star rating system. Unfortunately no information was available in the summary as to how these hotel types were defined.

In this case the KPI used was m^3/guest/year, using occupancy as the normalising factor rather than bedspace. Information on guest nights (real occupancy) was available for the audited hotels. However, for the benchmarking exercise an assumption was made for occupancy based on Wales Tourist Board figures. The authors of the Welsh Hotels study note that a benchmark per guest night is a more accurate KPI, although they concede that information is difficult to obtain and measure.

The eight hotels audited all fell into the "high" or "very high" water use category and although all the hotels showed improvement after the fitting of water efficient equipment, only one moved into the "average" category. None were in the good or best practice categories, even after efficient equipment had been installed. The summary did not contain information on the thresholds of these benchmarks and cannot be compared with the analysis results for the current project. As the more detailed reports for the project were not published, the project team were unable to determine why those hotels which had been improved did not reach the "good" or "best practice" categories. However a number of case studies were produced[7], which are summarised and the findings utilised in the guidance document that accompanies this report.

4.1.3 International Hotels Environment Initiative (IHEI)

This group of major hoteliers operate a benchmarking website called BenchmarkHotel[8]. The website requires a subscription upon which the hotelier can input detailed information about their hotel(s) including leisure centres, laundry and maintenance schedules. The hotel is then benchmarked in the database against other information input by users depending on the facilities and methods of management (ie outsourcing of laundry etc). The information is much more detailed, designed for larger prestige hotels and is populated by larger chains of hotels that have the ability to obtain much of this data from in-house Building Management Systems (BMS). The operators of the website admit that they have had little success in attracting smaller hoteliers to use the benchmark and have little information on the smaller premises. As with the Welsh Hotels study, the benchmarks are organised in terms of water consumption per guest night, and the hotelier is required to input this information into the program. Due to the nature of such benchmarking operations, the benchmark thresholds are unavailable unless a subscription is made and data to improve the data set is input. The information is not widely available or published.

4.2 Defining a KPI

Literature available for hotel water consumption studies (as above) was considered in proposing a key performance indicator for this sector. Following discussions with

BenchmarkHotel, it was clear that the larger hotels are already aware of the need to monitor and target the water consumption of their premises and many are already doing so. Discussions with Hilton Hotels showed that they have a very sophisticated system of monitoring involving their Building Management System (BMS) and have already made progress in this area. Many of the other large chain hotels are reportedly following similar strategies. Subsequently the project team proposed limiting the current study to smaller hotels only. The success of BenchmarkHotel among larger hotels would also support the assumption that these larger premises already have adequate access to benchmarking information. This proposal was agreed by the project steering group early in the project.

In defining a KPI for the hotels sector, the following factors were thought to have an influence on hotel water consumption:

- occupancy
- level of equipment/facilities present in the hotel
- age of property (or latest refurbishment)
- type of sanitaryware installed
- maintenance and management behaviour
- guest/user behaviour.

Ideally a KPI will eliminate as many of these variables as possible by normalising them into the calculation, leaving a minimal number of variables upon which the benchmark can vary. The extent to which this can be done depends on the availability of useful data, the KPI should be proposed with the data availability in mind.

The study was desk based and therefore information on the type of sanitaryware installed and the behaviour of management and maintenance staff and guests could not be identified. This was only identified from audits of each hotel, which on a study of this scale was not feasible.

Of the remaining variables, it was found to be impossible to identify the age of property or latest refurbishment date without contacting each hotel – again this was unfeasible.

It was possible to estimate the level of equipment and facilities available in each hotel using the familiar "star" rating used by the British Tourist Board (now Visit Britain – the British Tourist Authority). Although this is not an accurate indicator of more "luxurious" facilities (although it can highlight the presence of disabled access or other unconnected factors), it does provide an indicator of the level of facilities that a hotel may offer such as power showers, air conditioning and so on.

The following is a brief outline of the star ratings:

- one star establishments are practical accommodation with a limited range of facilities and services. There may be a restaurant or other eating area open to residents for breakfast and dinner. There may also be a bar or facility to serve alcoholic drinks
- two star establishments will have all of the one star requirements but will be more comfortable and better equipped
- three star establishments will have all of the facilities that one and two star establishments have, but in addition will offer a significantly greater quality and range of facilities and services. They can be more spacious and will offer a more formal style of service with a receptionist on duty, room service and a laundry service

- four star establishments have everything one, two and three star establishments have. They offer superior comfort and quality. All bedrooms will be en-suite. The hotel will have spacious public areas with a strong emphasis on food and drink. Room service should be available 24 hours a day

- five star establishments have everything one, two, three and four star establishments have. They are spacious, luxurious establishments of the highest international quality throughout. There will be a range of extra facilities and the hotel will be among the best in the industry.

The "star" rating is a voluntary standard and therefore not all hotels have such a rating. Those that are "no star" establishments should not be assumed to be small hotels without the facilities attributable to a one star establishment – they may equally be very well equipped but for whatever reason do not choose to participate in the star rating system. Similarly the star rating system reflects the presence of other facilities in a hotel that may not affect water use, such as disabled access.

Occupancy was thought to be the most important variable in determining water consumption in a hotel. This is supported by the other studies outlined above. However, occupancy is very difficult to determine, particularly in smaller hotels. Calls to a random sample of hotels indicated that guest numbers over a period of time are unrecorded other than in booking details. In order to recover this occupancy information a search through booking forms etc would be required. This was not possible for the current study, and in discussion with the project steering group, it was thought unlikely that the majority of small hoteliers would take the time to produce and tabulate such information in order to benchmark their water consumption, as this would be a time consuming exercise.

Another option would be to use standard occupancy information from the Tourist Boards to make assumptions for hotels in England, Wales and Scotland. This was the approach taken for the Welsh Hotels study. However the wide variation in occupancy likely across the large region covered by this study made this option less appropriate.

It was proposed that the KPI should use bedspaces as a proxy for occupancy in hotels. Bedspaces are the number of beds available in a hotel (for example, a double room would comprise two bedspaces, and a triple room would comprise three) which although not as accurate as occupancy, should provide a benchmark that can be easily calculated by all hoteliers without excessive collection of additional information. This indicator is likely to produce less accurate results for certain hotels. For example, some only have double rooms, promoting single occupancy particularly in areas frequented by business travellers.

In order to test the applicability of bedspace, a very small sample was analysed from a single town in Berkshire where business and tourism are equally likely. This minor analysis illustrated that the relationship between water consumption and bedspace was strong, and the project team concluded that this was an appropriate KPI.

The KPI for hotels was therefore selected as **cubic metres of water per bedspace per annum**.

4.3 Other water consumption variables

There are a number of other variables that may be expected to influence hotel water consumption. These include:

- presence of swimming pools and/or leisure centres
- presence of large grounds or gardens that may require irrigation
- presence of restaurants serving a significant number of covers
- presence of commercial laundry facilities.

Of these, it was possible to identify whether the hotel has a swimming pool from readily available tourist guides to hotels, but information was inconsistent for the other variables. Investigation of the relationship between these variables and water consumption may indicate certain "influencing factors" that could be used to assist the benchmarking process.

4.4 Obtaining data

In order to ensure a statistically robust sample, the project team agreed that approximately 0.5 per cent (and preferably 1 per cent) of the total number of hotel buildings in the UK should be obtained. Based on information from the British Tourism Authority, this was suggested to be a minimum of 300 hotels, and preferably 600.

A number of options were considered for obtaining information on hotel accommodation, address and facilities. One difficulty was obtaining a randomised source of information from across the UK, as often such databases are regionally held. However a source was obtained to provide information on approximately 3500 hotels across the UK covering address data, star rating and whether the hotel has a swimming pool.

Data was obtained from:

- British Tourist Board (star ratings)
- hotels database provider through British Tourist Board
- water companies (water consumption matched to addresses provided).

Unfortunately the data does not include information on whether there are restaurant facilities, laundry facilities or gardens. Additionally, some hotels included self-catering facilities which was not explicitly included in the data obtained. It was determined that this was the best available data and was possibly purchased. An additional risk was identified that the accuracy of the data could not easily be verified without contacting each hotel. Further information on the anomalies found is discussed below.

4.4.1 Separating data into water company areas

Hotel information was separated into the relevant water company areas by the postcodes (by using a database routine). The number allocated to each water company is as follows:

Table 4.2　Water companies and hotels allocated to each

Company	Number of hotels
Anglian Water	169
Bournemouth and West Hampshire Water	14
Essex and Suffolk Water	24
Scottish Water	398
South West Water	410
Southern Water	26
Thames Water	115
Three Valleys Water	13
United Utilities Water	370
Welsh Water	117
Total	**1656**

Approximately 1850 hotel addresses remained that are either not within one of the participating water company's areas, or for whom the company responsible cannot be identified. However sufficient records were available to allow the analysis to progress once matched water consumption information was obtained.

Water use data was obtained from water companies to match the data for the years 2001 and 2002. Following sorting of this data to remove null or duplicate records, or those for which insufficient information was available, 403 records remained for analysis. Although less than hoped, this is greater than the minimum number of records required (300) for a robust sample.

4.5 Hotel data analysis

On initial inspection it was found that the original data had a number of errors. The most critical error, for the purposes of the analysis, was in the distribution of room type. In order to determine the number of bedspaces, the number of people for each type of room was multiplied by the number of each type (ie two bedspaces for a double or twin, one bedspace for a single). For some of the hotels, the split provided by the database supplier did not always match the total number of rooms in the hotel. Further investigation of those hotels where this proved to be an issue was undertaken on the internet and where possible, a corrected room split was recorded. While undertaking this analysis it was also determined that there may be other anomalies in the data. These included some hotels which had self-catering accommodation and hotels which had closed.

A further difficulty arose in determining the number of bedspaces in family rooms. Some hotels have family rooms for three people and some for four. Some sensitivity analysis was undertaken as to the potential effect on the data of three or four people. In addition, the research contractor investigated a random sample of hotels to determine the most common number of bedspaces in family rooms. Initial analysis suggested that there was little effect on the data and that the majority of hotels in the random sample had three bedspaces in family rooms. Without further, more robust data, it has been assumed that family rooms have three bedspaces. However the effect on the analysis is minimal as the number of family rooms in each hotel was small in proportion to doubles, twins and singles.

Following removal of those hotels where the number of bedrooms did not correlate to the totals provided by the database provider, the hotels with zero consumption were removed and the remainder were plotted on a graph. The analysis was completed on 279 hotels from the initial sample of 1656.

Table 4.3   Hotel sample sizes

Hotel rating	Swimming pool	Number of hotels in category
1 star	Yes	0
1 star	No	7
2 or 3 star	Yes	18
2 or 3 star	No	94
4 or 5 star	Yes	8
4 or 5 star	No	56
No rating	Yes	11
No rating	No	85
		279

A preliminary analysis charted the simple water consumption during each year 2001 and 2002 against bedspace, to investigate the correlation between the two years. In 2001 the foot and mouth outbreak in the UK resulted in a reduction in tourism within certain parts of the UK. It also enabled investigation of the correlation of the data to bedspace, giving a preliminary indication of whether the data would produce useful benchmarks or not.

4.5.1 Initial correlation and aggregated analysis

The analysis showed that in both 2001 and 2002 there is a very significant grouping of data below the 200 m^3/bedspace mark. Although there are small differences between the two years, this definite grouping indicated that further analysis may produce appropriate benchmarks. It also indicated that the foot and mouth outbreak did not have a significant effect on the hotels surveyed within the two year period considered. The data was grouped toward lower consumption, with a small group of hotels with much higher consumption evident. Further analysis would be required to ascertain if these were outliers, or if they had higher consumption for a reason that could be identified.

The data was sorted into ascending order of water consumption per bedspace and a trendline added. The R^2 value for a linear trendline for 2001 is 0.48 and for 2002 is 0.54. The correlation of the data for 2001 and 2002 is evidently similar. However due to the foot and mouth outbreak in 2001 and due to the slightly better R^2 value for 2002 (implying a better correlated data set), the project team decided to focus on the data for water consumption in 2002.

Figure 4.1 shows an enlargement of the relevant section of the frequency distribution chart illustrating the 25, 50 and 75 percentiles. These figures are for the whole data set and represent an aggregated result.

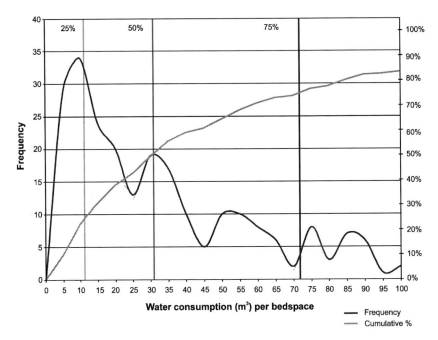

Figure 4.1 Chart showing frequency distribution of water consumption per bedspace for consumption in 2002

The initial water consumption analysis indicates that a typical usage (50 per cent) of the hotels analysed would be approximately 31m³/annum, with approximately 72m³ per annum being indicative of "excessive" use. On discussion with the water company members of the project steering group, this was felt to be a little low. Only the higher 75 per cent figure was broadly what the water companies would expect.

If these figures are converted to litres per day (assuming 365 operating days per year), they equate as follows:

Table 4.4 Overall hotel water usage

	m³/annum	Litres/day
Best practice	12	33
Typical	31	85
Excessive	72	197

Note: Figures have been rounded for ease of use

There is another possible explanation of the figures that is causing them to be unexpectedly low – potentially another factor that the aggregate analysis cannot express. Indeed, the graph shows a number of peaks in the data, which is indicative of specific factors affecting the water consumption. In order to determine the influencing factors, further analysis has been undertaken based on the information in the data set.

The information in the data set includes the "star rating" of each hotel and whether there is a swimming pool. Information is unavailable as to whether each hotel has a garden, for example, which could have a large influence on water consumption.

4.5.2 Dividing hotels into types for further analysis

The initial investigation was to split the data into those with swimming pools and those without. The frequency distributions are shown below in Figures 4.2 and 4.3.

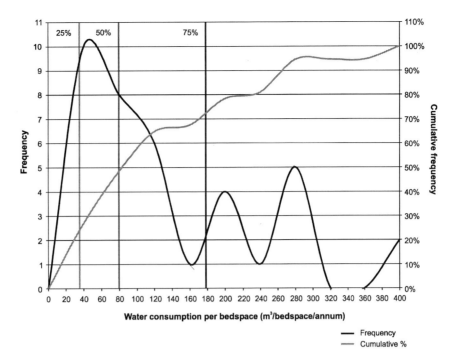

Figure 4.2 Chart showing frequency distribution of water consumption for hotels with swimming pools for consumption in 2002

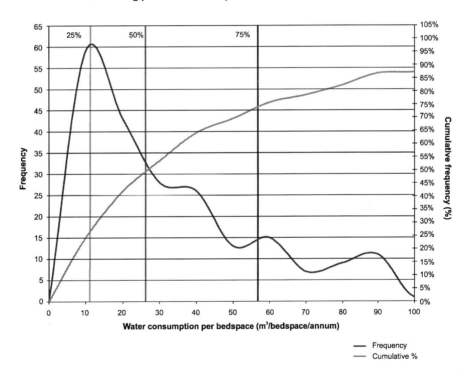

Figure 4.3 Chart showing the frequency distribution of water consumption per bedspace for those hotels without swimming pools for 2002

Figure 4.2 indicates a number of peaks in the data suggesting that having a swimming pool is not the only influencing factor. Figure 4.3 is a larger dataset and achieves a smoother curve suggesting that hotels without swimming pools have fewer influencing factors. This also suggests that there may be specific influencing factors in hotels with swimming pools.

This could be that most hotels with swimming pools may also have gardens, or that where there is a swimming pool, the residents have more showers and baths than in hotels without swimming pools. A further factor may be that the swimming pool and changing areas could be associated with fitness centres of gyms, with showers and other equipment open to non-residents. This would mean that the occupancy factors are not relevant. However none of these can be ascertained and it is not appropriate to speculate without sufficient data.

4.6 Hotel water consumption by star rating

Initially the hotels were analysed by "star" rating to determine whether this resulted in a better correlation than the presence or absence of a swimming pool.

The hotels were divided into one, two or three star establishments, and four or five star establishments as shown in the two charts below.

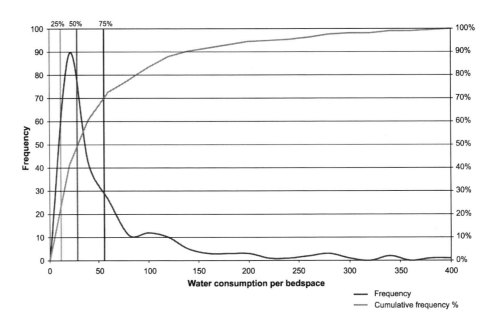

Figure 4.4   Water consumption of hotels up to three star rating

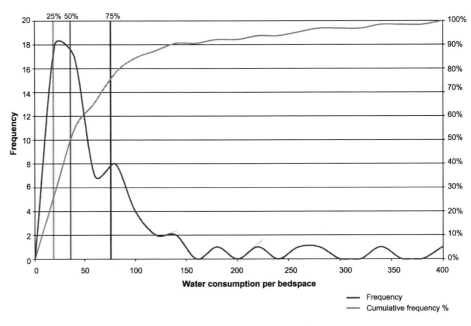

Figure 4.5   Water consumption of four and five star hotels

Figures 4.4 and 4.5 show the frequency distributions of water consumption for hotels up to three star and for four and five star hotels. They clearly illustrate that the "star" rating of the hotel is not the only factor influencing the water consumption of the hotels, due to the "noise" in the data. The number of peaks and the spread of data would indicate that there are other factors to be discovered.

In order to investigate this, the star rating of the hotel and the presence or absence of a swimming pool were combined in a further analysis.

4.7 Hotel water consumption by star rating and presence of swimming pool

In order to divide the hotels into manageable categories for analysis, they were distributed into categories depending on their star rating.

The categories were split as follows (an explanation of the star rating system can be found in Section 4.2):

Category 1: One star rated establishments.

Category 2: Two or three star rated establishments.

Category 3: Four or five star rated establishments.

Other: Undefined establishments, those which do not participate in the "star" rating system.

Figure 4.6 shows an enlargement of the frequency distribution per bedspace for those hotels with no rating. This also indicates the 25, 50 and 75 percentile.

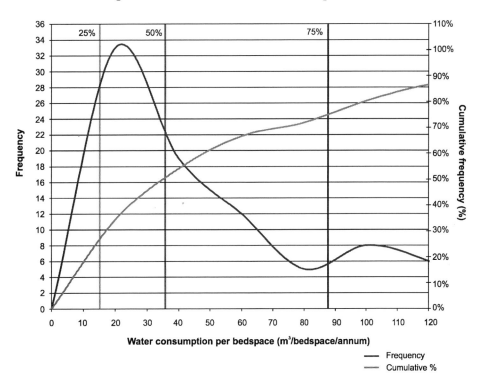

Figure 4.6 Chart showing frequency distribution of water consumption per bedspace for "other" category hotels

By comparing Figure 4.6 for the hotels with no rating to the hotels allocated in categories one, two and three below, it appears that there is some discrepancy in the hotels with no rating. There may be various reasons for this – hotels with no rating may be pubs or bed and breakfasts, where there may be a high level of non-resident water use, for example. This was found in the Welsh Hotels study, but without sending questionnaires to all the hotels in the list, there is no way of telling from the data.

The chart shows that although the water consumption is variable, as might be expected with such a potentially wide variety of hotel types, there is a clear grouping at the lower consumption end of the scale. The median point (50 per cent) falls at approximately 35 m^3/bedspace/year (96 litres/day), with the first quartile (25 per cent of hotels have better water consumption) falling at approximately 18 m^3/bedspace/year (49 litres/day).

The chart showing category one hotels (Figure 4.7) by comparison has a median point (50 per cent) at only 9.5 m^3/bedspace/year (26 litres/day) and the first quartile (25 per cent) at only 7.5 m^3/bedspace/year (21 litres/day). The graph shows a clear grouping of hotels at the lower consumption part of the chart, with many fewer hotels consuming larger volumes of water.

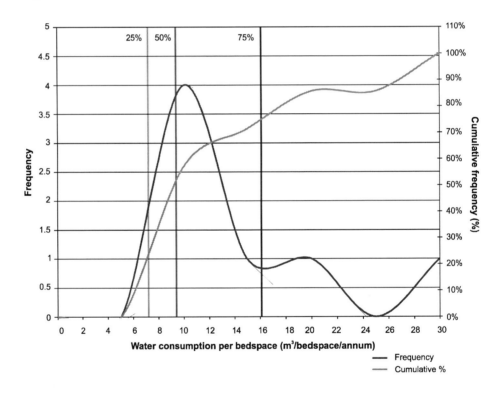

Figure 4.7   Frequency distribution of water consumption (m^3) per bedspace for category one hotels

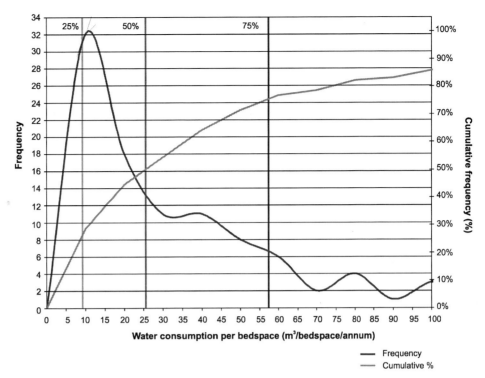

Figure 4.8 Frequency distribution of water consumption (m³) per bedspace for category two hotels

The water consumption for the category two hotels as shown by Figure 4.8 is less variable, but again there is a clear grouping toward the lower consumption end of the scale.

In fact this grouping is so pronounced that the benchmarks for this category of hotel are similar to those for the category one hotels. The median point (typical practice) is in this case at 25 m³/bedspace/year (69 litres/day), with the first quartile (best practice) point at approximately 10 m³/bedspace/year (27 litres/day). This was unexpected, but on reflection may be due to the increased likelihood of this category of hotel having more modern fittings and improved sanitaryware such as showers in each room rather than baths. This is broadly supported by the Welsh Hotels study although their benchmarks are not available. Further studies will be able to build on the analysis in order to prove or disprove this working hypothesis.

The Welsh Hotels study also found that at least in some category one establishments, there was a likelihood of increased non-guest usage of water (such as the establishment also operating as a restaurant or pub) which would significantly increase water use. This would often occur in the more rural areas that were considered by the Welsh study, and may account for the differences between the Thames Water study (higher proportion of municipal area hotels, given that it covers the highly populated south east of the UK) and the current research.

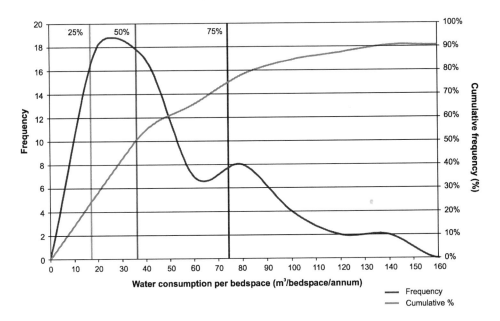

Figure 4.9 Frequency distribution of water consumption (m³) per bedspace for category three hotels.

The chart shows that this group of hotels had the greatest variation in water consumption of any group. In part this is due to the sample being smaller, as the research concentrated on smaller premises to avoid overlap with the BenchmarkHotels project. However a great variation is due to the widely varying facilities present in this category of hotel.

In this case, the median point (typical practice) is at 38 m³/bedspace/year (104 litres/day) with the best practice (25 per cent) point at 18 m³/bedspace/year (49 litres/day). Once again this is considerably different to the findings of the Thames Water study, where benchmarks at more than double this level were recommended.

4.8 Hotels with swimming pools by category

An analysis of the potential "influencing factor" associated with swimming pools was undertaken. In each case, the category of hotel is divided into those with and without pools, and graphs drawn for each to allow comparison of the differences made. There are no hotels in category one with a swimming pool, so no figure for this category is available.

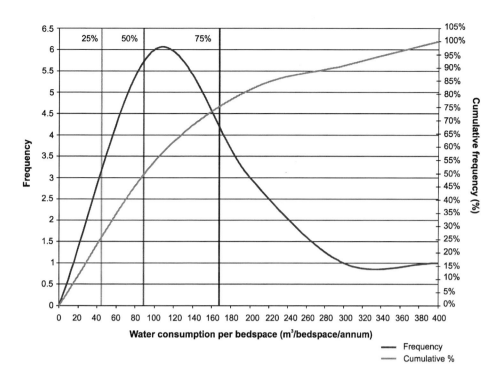

Figure 4.10 Frequency distribution of water consumption (m³) per bedspace of hotels with no stars and with swimming pools

Figure 4.10 represents the broadest range of hotel types, as these are undefined hotels, but only assesses those with pools. In this case the median point (50 per cent) is at 90 m³/bedspace/year (247 litres/day) and the lowest quartile (25 per cent) at 41 m³/bedspace/year (112 litres/day).

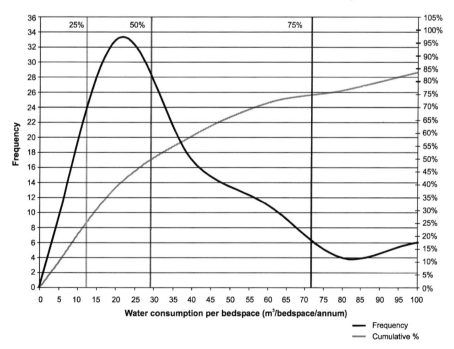

Figure 4.11 Chart showing frequency distribution for water consumption per bedspace for hotels with no star and with no swimming pool for consumption in 2002

Again, Figure 4.11 represents the broadest range of hotel types, as these are undefined hotels, though in this case they do not have pools. In this case the median point (50 per cent) is at 12 m³/bedspace/year (33 litres/day) and the lowest quartile (25 per cent) at 29 m³/bedspace/year (80 litres/day).

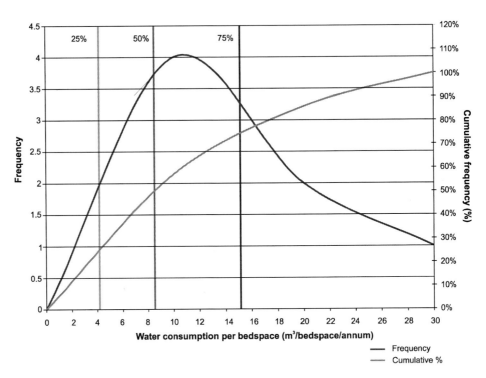

Figure 4.12 Chart showing frequency distribution for water consumption per bedspace for hotels with one star and no swimming pool for consumption in 2002

Figure 4.12 shows hotels with one star which do not have a swimming pool. In this case the median point (50 per cent) is at 9 m³/bedspace/year (25 litres/day) and the lowest quartile (25 per cent) at 4 m³/bedspace/year (11 litres/day).

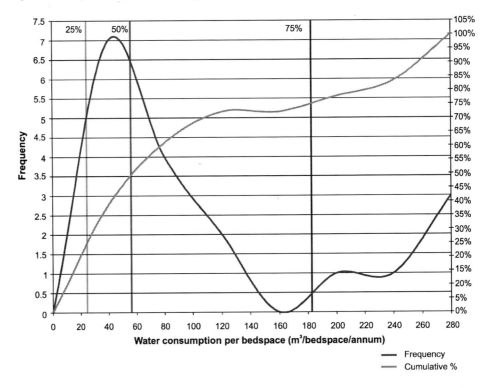

Figure 4.13 Chart showing the frequency distribution of water consumption in hotels with two or three stars with swimming pool for consumption in 2002

Figure 4.13 shows hotels with two or three stars which have a swimming pool. In this case the median point (50 per cent) is at 58 m³/bedspace/year (159 litres/day) and the lowest quartile (25 per cent) at 22 m³/bedspace/year (60 litres/day).

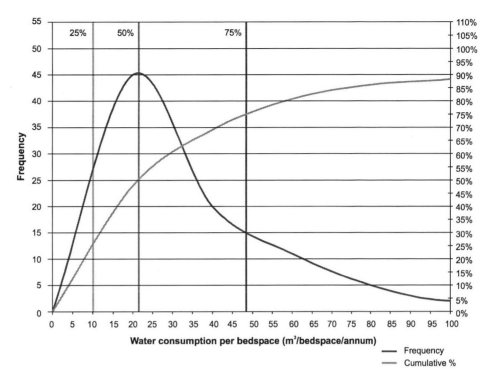

Figure 4.14 Chart showing frequency distribution of water consumption per bedspace for hotels with two or three stars without a swimming pool for consumption in 2002

Figure 4.14 shows hotels with two or three stars without a swimming pool. In this case the median point (50 per cent) is at 22 m³/bedspace/year (60 litres/day) and the lowest quartile (25 per cent) at 10 m³/bedspace/year (27 litres/day).

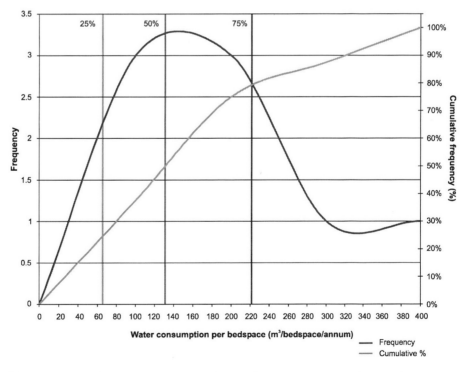

Figure 4.15 Chart showing frequency distribution of water consumption per bedspace for hotels with four or five stars and a swimming pool for consumption in 2002

Figure 4.15 shows hotels with four or five stars which have a swimming pool. In this case the median point (50 per cent) is at 130 m³/bedspace/year (356 litres/day) and the lowest quartile (25 per cent) at 62 m³/bedspace/year (170 litres/day).

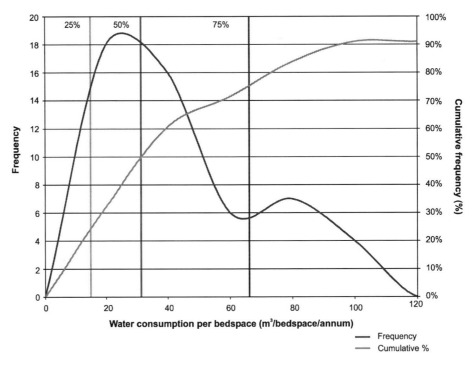

Figure 4.16 Chart showing frequency distribution of water consumption per bedspace for hotels with four or five stars and no swimming pool for consumption in 2002

Figure 4.16 represents hotels with four or five stars without a swimming pool. In this case the median point (50 per cent) is at 31 m³/bedspace/year (85 litres/day) and the lowest quartile (25 per cent) at 15 m³/bedspace/year (41 litres/day).

4.9 Recommended hotel benchmarks

Given the above analysis it is evident that the presence of a swimming pool has a marked influence on the water consumption of the hotel. For each category that had hotels with and without pools, the presence of a pool more than doubled the water consumption at each percentile level.

For this reason it is recommended that two separate benchmarks are produced, for those hotels with and without pools, as shown in the table below. The benchmarks have also been shown graphically in charts below the table, separated for those with and without pools.

Table 4.5 Benchmarking outputs for hotels

Category	Hotel rating	Swimming pool	Number of hotels in category	Benchmarks (m³/bedspace/annum)		
				25%	50%	75%
Cat 1	1 star	Yes	0	None		
Cat 1	1 star	No	7	4	8.5	15
Cat 2	2 or 3 star	Yes	18	25	55	183
Cat 2	2 or 3 star	No	94	10	22	48
Cat 3	4 or 5 star	Yes	8	65	130	222
Cat 3	4 or 5 star	No	56	15	31	66
Other	No rating	Yes	11	45	89	169
Other	No rating	No	85	12	29	72

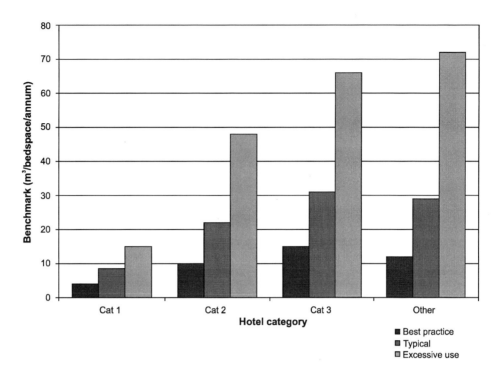

Figure 4.17 Chart showing benchmarks for each category of hotel without pools

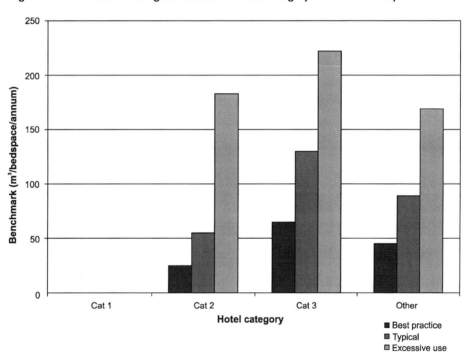

Figure 4.18 Chart showing benchmarks for each category of hotel with pools

It is evident from these charts the increase in water consumption for hotels with pools, however this significant difference was not reported in the Thames Water study, which had concluded that swimming pool use was not significant in water consumption benchmarking.

When comparing the benchmarks produced by this study and by the Thames Water study, they are in close agreement with the results for hotels with pools, as shown in Figure 4.19.

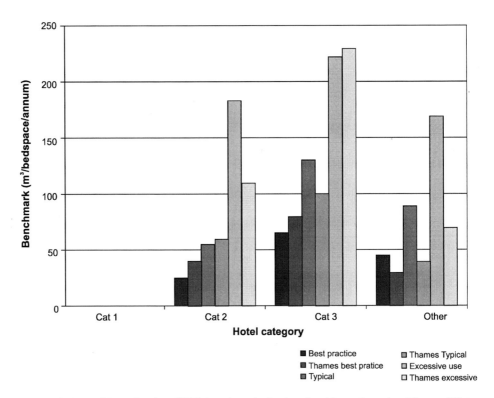

Figure 4.19 Chart showing CIRIA benchmarks for hotels with pools against Thames Water study benchmarks

It is not possible to identify which category of hotel investigated by Thames Water had pools from the information available, but the similarity between results would suggest that the current proposed benchmarks are suitable.

In order to create a benchmark for category one hotels with pools (as none were included in the analysis), the multipliers between the quartiles for category two hotels have been used as shown in the table below. The benchmarks have been split into those hotels with and without pools, and the figures obtained by analysis have been rounded to create easily accessible benchmarks.

Table 4.6 Recommended benchmarks for hotels with swimming pools

Category	Hotel rating	Benchmarks (m³/bedspace/annum)		
		Best practice	Typical	Above average
Cat 1	1 star	9	25	60
Cat 2	2 or 3 star	20	60	185
Cat 3	4 or 5 star	60	130	220
Cat 0	No rating	40	90	170

In order to provide an additional verification of the benchmarks, they were compared to other small sets of unpublished data provided by the water companies. Although meaningful statistical comparisons could not be undertaken due to the small size of the unpublished data set (all of which were within category two), it has been determined that the results are comparable.

The unpublished data shows water consumption between 16 m³/bedspace/year and 95 m³/bedspace/year for category two hotels with swimming pools. This is in line with the report findings and provides a good correlation.

Table 4.7   Recommended benchmarks for hotels without swimming pools

Category	Hotel rating	Benchmarks (m³/bedspace/annum)		
		Best practice	Typical	Above average
Cat 1	1 star	5	10	15
Cat 2	2 or 3 star	10	20	50
Cat 3	4 or 5 star	15	30	65
Cat 0	No rating	10	30	70

The unpublished data has water consumption of between 23 m³/bedspace/year and 81 m³/bedspace/year for category two hotels without pools – this is higher than the current analysis which had the 75 per cent at 50 m³/bedspace. However the majority of the data is at about 40-60 m³ which correlates with current analysis.

These benchmarks are proposed as appropriate for the hotels sector given the current research. Further analysis may well improve and refine the benchmarks and it is hoped will shed further light on other influencing factors within hotels that could not be investigated within the current study.

It is important to note that the benchmarks generated during this study have been produced to be independent of occupancy. This is because occupancy rates will vary widely across hotels, depending on the type of hotel, the location and their management and operation. Some hotels close in winter months, for example, and will have lower occupancy across the year, although they may be fully occupied in the summer months. Some hotels are fairly under occupied during the week but will be very busy at weekends. A hotel that is fully occupied all year is likely to have higher water use than the benchmark. To make the benchmarks more accurate, they could be normalised regionally using figures obtained from the British Tourist Authority, and it would be beneficial for water companies who intend on using this data to produce regional information for those hotels.

5 Benchmarking offices

5.1 Water use in offices

There are approximately 278 000 commercial offices in the UK[3], and despite this sector being the target of many guidance documents and benchmarks concerning energy use, there is little or no information helpful on water benchmarking. In part this is due to the particular commercial arrangements associated with office buildings, where many have only one meter but several tenants, or where water is not metered at all.

Nevertheless the success and usefulness of energy benchmarks led CIRIA to conclude that investigation of office water use would yield information useful to tenants, building operators and owners.

The approach for investigating this sector was based on previous work carried out by other research teams (information of these sources follows) and on discussions with the project steering group. It was necessary to consider not only the information that would be of most use to the offices sector, but also the information that it was possible to obtain within the scope of the study. Further discussion of the use of information and selection of benchmarks and KPIs is provided below.

5.1.1 Thames Water

Thames Water commissioned a study, completed in August 2000, into the water use in offices within their region which not only considered the total water use consumed by this sector in the Thames Water region, but suggested benchmarks and potential reductions in water consumption that may be possible to achieve. The study[10] collected data on floor area and water consumption from various sources and used this to investigate the differences in water consumption from offices of differing sizes. As a result, the authors recommended targets for offices in two bands depending on size, with a third band for offices with restaurants. The authors recognised that employee number should in theory be a closer match to water consumption than floor area but they were unable to ascertain the number of employees. They therefore calculated a target for each band derived from mean water consumption per square metre of office area, and assumed average occupancy numbers based on the British Council for Offices Guide 2000, of 15 m² per person.

The targets provided are equivalent to the 25 percentile (first quartile) and would represent a "good practice" target:

Table 5.1 Thames Water study outputs for offices

	m³ water per annum per m² net internal area	m³ water per annum per employee
Offices <1000m² net internal area	0.29	4.4
Offices >1000m² net internal area	0.46	6.8
Offices with restaurants, add:	0.1	1.5

5.1.2 Government estate

The Government Estate makes up a significant proportion of non-domestic space in the UK. The Government is also committed to demonstrating a move toward more sustainable management of their space in line with commitments to sustainable development. The current target is set at 7.7 m³ per person per year by 31 March 2004 for existing buildings and 7 m³ per person per year for all new buildings and major refurbishments where design commences after 2002[10]. "Per person" is not defined therefore it is assumed that the document means "per occupant", not "per employee". The Environment Agency's Waterwise guide[7] uses the same benchmark for existing properties. It appears that these benchmarks were not derived directly from a measured data analysis, but by calculation and it is likely that they will be revised as more data becomes available.

5.1.3 Watermark

Watermark is a web-based benchmarking tool for publicly owned buildings originally set up by OGC Buying Solutions. It is available for any user to input their own water consumption that is then automatically benchmarked against other inputs from users, although its primary target and users are government buildings. An analysis of the inputs from the office sector was completed in May 2003[12] and a typical, best practice benchmark was calculated. The statistical analysis indicated that occupancy was the primary driver in water consumption in the offices. Although other drivers including floor area, opening hours and building age were considered, they had a very low correlation. Benchmarks were produced per full time employee per year based on information provided by facility managers.

The benchmarks recommended are 9.3m³/person/year for a typical office, and 6.4 m³/person/year as a best practice target.

5.2 Defining a key performance indicator

The KPI for offices was not straightforward. The primary factors thought to have an influence on office water use (based on the above studies) would be:

- occupancy
- size of office (if occupancy is unavailable)
- age of property (or latest refurbishment)
- type of systems and fittings installed
- maintenance and management behaviour
- user/employee behaviour.

Without a full audit of each office, it was unfeasible to find information on the age of the property or type of sanitaryware installed, however it was possible to draw some conclusions based on other studies (see Section 5.3). This meant that any investigation would need to be based on either occupancy or floor area.

As the investigation of previous studies above shows, they have used each of these factors to define benchmarks, depending on the data available and that proving most robust.

The risk of information for either of these factors being unavailable was identified at an early stage, and it was decided to carry out a limited preliminary exercise to investigate the correlation between office area and occupancy and whether a conversion factor could be identified.

This analysis carried out on a limited sample of offices (from a single employer) for which occupancy and area could be verified, showed that there was a strong correlation between the two factors. This gave the project team confidence that should only one form of data be available, then conversion would be possible whichever benchmark was decided upon.

The project team had initially considered that a KPI based on floor area would be most appropriate. This conclusion was reached following previous studies (though unrelated to water consumption) dealing with commercial office areas and commercial property managers, who seldom knew the occupancy of their office. It is certainly true that while occupancy varies over time, the floor area is a more stable measurement. Finally it was thought that occupancy is a difficult factor to investigate since it can be interpreted in different ways – for example, total number of employees, or full time equivalent staff, or actual occupancy over a given period. An increase in flexible working practices also makes this more complex (a further discussion of this issue is included in Section 5.6.2).

Previous studies have shown that occupancy is a better correlation for water use and is used more often in benchmarking (such as by the government estate). Therefore in order to build upon previous research and add further value to the total data set available it would be best to utilise occupancy.

It was eventually determined that the project team would attempt to produce benchmarks based on both occupancy and floor area, although it was recognised that a single data set may not to be able to provide both pieces of information.

The key performance indicator for offices was determined to be **water consumption in m3 per person per year**, and **water consumption in m^3 per square metre per year**.

The client team also felt that it would be useful to provide the same indicator converted into **water consumption per litres per day** (assuming 253 business days per year). The user can then utilise whichever benchmark suits their circumstances.

5.3 Other water consumption variables

A number of fittings and systems that can be installed in some offices may influence water consumption in offices, including the following:

- presence of catering facilities including restaurants
- type of sanitaryware including showers
- use of wet air conditioning systems
- external managed space including grounds.

Currently there is no database of such information and, without obtaining considerable additional information directly from building owners or operators, it is impossible to make sensible assumptions about these variables since the nature of UK office buildings is extremely variable.

The Watermark benchmarking report[12] investigated this information, as their users were asked to input a greater volume of more detailed information. Where data on occupancy or floor area was missing, the analysis excluded their data. It is expected that the data analysed is more robust and accurate entered by users.

The Watermark investigation showed that there was no significant correlation for catering facilities, and this is unlikely to be significant factor in water consumption (unless very large). The type of sanitaryware is usually related to the age of the building unless significant refurbishment has been completed. However the investigation showed no significant correlation for the age of a building, suggesting that this is not a suitable indicator for water consumption. In order to investigate more closely, the Watermark team also considered the presence of showers, but these were also found to have no significant correlation with water consumption.

There was insufficient information concerning external space to allow analysis of this variable to take place, but as the majority of commercial offices will have little planted area (much of their external space will be car parking) this is thought to have a restricted influence for the majority of offices.

The primary drivers are likely to be occupancy and/or floor area which supports their use as a KPI.

5.4 Obtaining data

In order to ensure a statistically robust sample, the project team agreed that approximately 0.5 per cent (and preferably 1 per cent) of the total number of office buildings in the UK should be obtained. Based on information from BRE[3] this was suggested to be a minimum of 1390 offices, and preferably 2780.

Initially it was intended to obtain the information on addresses, floor area and occupancy from the project team's contacts within the UK commercial property market. This was successful in that close to 1000 office addresses and floor area information were received. However few of the property managers could provide information on occupancy, as this is determined by the company occupying the space – potentially more than one company. More data was also required to obtain a statistically significant data set.

A number of existing databases of information were considered, including the Estates Gazette (covering only London, Manchester and Leeds) and property management company databases. The most promising was information produced by the Valuation Office which includes both address and floor area information. At the time of this investigation however, the information was not in the public domain and could only be supplied in separate data sets, without any connection between area and address rendering it unsuitable for analysis. Since the completion of this project, the Valuation Office has now provided this information on their website, although only manually extractable. A future project could be to purchase a randomised sample of only office area from the valuation office and repeat the analysis based on this data.

The eventual option chosen was to purchase a randomised data set from the National Business Database, who record information on office address and occupancy across the UK. There was a concern that this database would provide information on offices that were in multi-tenanted buildings. This would considerably distort the results since the number of employees of that company would be less than expected compared with the water consumption for the whole building. In order to avoid this possibility, it was decided that only addresses for offices with greater than 350 employees would be obtained. Although there was still some risk, this would be minimised by selecting larger employers who are liable to occupy entire buildings. Accordingly 3000 addresses were purchased of which 2592 were found to be suitable following initial analysis.

These were sent to water companies for data matching with water consumption as shown in the table below:

Table 5.2   Water companies and offices allocated to each

Water company	Offices
Anglian Water	396
Bournemouth and West Hampshire Water	9
Essex and Suffolk Water	103
Scottish Water	469
South West Water	40
Southern Water	114
Thames Water	720
Three Valleys Water	174
United Utilities Water	309
Welsh Water	149
Northumbrian Water	109
TOTAL	**2592**

Unfortunately due to complications in obtaining suitable data (see Sections 3.2 and 3.6), only 222 records of matched water consumption data for the years 2002 and 2003 remained following the preliminary data cleaning and sorting phase.

Despite the low number of records, it was decided to proceed with the analysis, documenting the methodology to allow future researchers access to better data to supplement the information and add to the data set.

5.5   Office data analysis (floor area)

Initial analysis of the water consumption data against floor area found that although there were a low number of evident outliers that did not fit, the majority of results are grouped below 1 m^3/m^2. This is anticipated given the benchmarks found by other studies and provided a high degree of confidence that the data would provide useful information.

A process of manually removing the outliers and evident anomalous records was undertaken. There were two main reason for removal of outliers, either the water data provided was inaccurate (nil or widely variable readings), or the building was evidently not a true office when the address was examined. The accompanying CD-Rom to this project provides further information.

Once this process was complete a simple linear regression could be performed on the remaining data. The R^2 value for the regression line produced is 0.9875, which shows that the data correlates well with little deviation. This means that there is a strong relationship between office water use and floor area, for the data analysed.

The analysis showed that there was no significant difference between larger and smaller offices (with a range of between 30 m^2 and 6604 m^2) so the same benchmark and analysis can be used for all sizes of building. It is possible to plot the frequency distribution to identify quartiles of data for use in benchmarking.

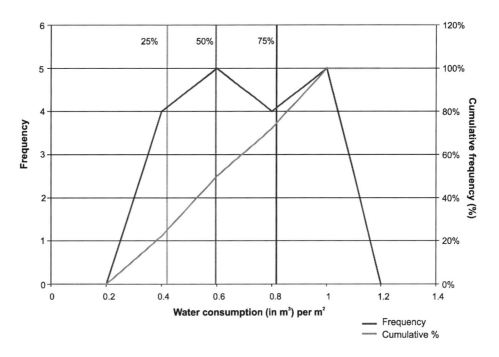

Figure 5.1   Chart showing frequency distribution for consumption by floor area

This chart divides the data into quartiles to establish a median point (50 per cent) that might be considered the typical water consumption of an office. It also shows the first quartile (the point at which 25 per cent of offices had lower water consumption) that would be considered the best practice target for office buildings, and the last quartile (75 per cent) which represents "excessive" water use. This is in line with the targets as agreed with the project steering group (see Section 5.2).

Based upon this analysis, the proposed benchmarks (rounded for ease of use) would be:

Table 5.3   Recommended benchmarks for offices by floor area

	Cubic metres per year	Litres per day (assuming 253 days per business year)
Typical use	0.6m³/m²/annum	2.4 litres/m²/day
Best practice use	0.4m³/m²/annum	1.6 litres/m²/day
Excessive use	0.8m³/m²/annum	3.2 litres/m²/day

This best practice figure closely matches the target provided for larger office buildings (also based upon the first quartile) of David Bartholomew Associates' Thames Water study[9] of 0.46 m³/m²/annum.

5.6   Office data analysis (occupancy)

The project team also wished to investigate occupancy, as this was envisaged to be a closer match for water use than floor area. A dataset of office address and occupancy was purchased from a national business database and matched to water consumption for the years 2002 and 2003. These were obtained from questionnaires supplied by the database company who asked companies to fill in their employee numbers. It is likely that unless the questionnaire was very descriptive, the facility managers would have completed it using **total employee numbers**.

The initial analysis of water consumption data was performed in order to chart simple water consumption against occupancy. This was to enable the project team to determine whether the data appeared sufficiently robust enabling further analysis. This analysis showed that the data is reasonably well clustered with few outliers, and preliminary analysis demonstrated that the data would be suitable for further analysis

Again it was necessary to remove evident outliers which involved manual examination of the data. There were several instances where some of the water consumption data appeared to be inaccurate (null or evident inconsistencies between the two years) and had to be removed from the data set. Also there were a number of instances where the record was not a true office and may include water used for other purposes. These were also removed, and a linear regression performed on the remainder of the data. Details of the records removed and the reasons for this are contained on the accompanying data CD.

When these anomalous readings were removed it was possible to conduct a regression analysis for the remaining data. The correlation is shown to be very good by the R^2 value which in this case is 0.969. This is not quite as good as the correlation in the data using floor area, but this is to be expected as a a bigger sample was used.

As with the previous offices analysis, there was no significant difference found between the offices with more or less employees (the range of occupancy was 356 employees to 6410 employees) therefore there was no requirement to provide a separate benchmark for different size offices. It is possible to divide this data into quartiles in order to propose benchmarks.

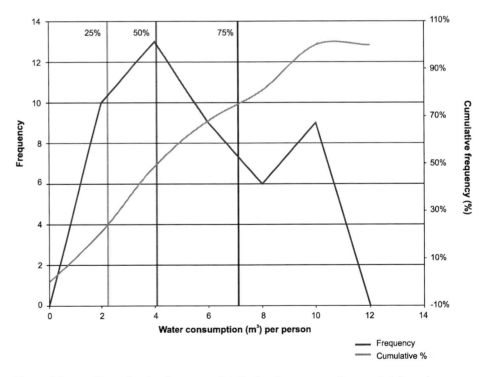

Figure 5.2   **Chart showing frequency distribution for consumption per total employee**

This chart divides the data into quartiles to establish the median point (50 per cent) that might be considered the typical water consumption of an office (when calculated on a per employee basis). It also shows the first quartile (25 per cent) which would represent a best practice target and the last quartile (75 per cent) which would represent excessive water use. The use of these quartiles as benchmarks was agreed with the project steering group and in line with other research studies (see Section 5.2).

Based upon this analysis, the proposed benchmarks would be:

Table 5.4   Recommended benchmarks for offices by employee numbers

	Cubic metres per year	Litres per day (assuming 253 days per business year)
Typical use	4.0m³/employee/annum	15.8 litres/employee/day
Best practice use	2.0m³/employee/annum	7.9 litres/employee/day
Excessive use	7.0m³/employee/annum	27.7 litres/employee/day

Note: Figure is rounded for ease of use

A comparison is shown in the table below using the typical and good practice benchmarks. The "excessive" benchmark is not compared as this was not calculated by the other studies.

Table 5.5   Table showing comparison of proposed benchmark and other research benchmarks

	"Best" or "good" practice target	"Typical" practice benchmark
David Bartholomew Assoc Thames Water study*	4.4m³/employee/year (offices <1000m²) 6.8m³/employee/year (offices > 1000m²)	
Government Estate		7.7m³/person/year
Watermark	6.4m³/full time employee/year	9.3m³/full time employee/year
Current research	2.0m³/employee/year	4.0m³/employee/year

* Note that the Thames Water benchmarks per employee were **calculated** from the office floor area data and do not represent true employee figures at these offices, therefore they are not an ideal comparator against measured data.

It is evident from this table that the benchmarks identified by the current research are lower than those previously proposed. There are many potential reasons for this and not all can be investigated.

However it is the opinion of the project team that there are two major factors at work in this calculation:

- the method of calculating the figure for personnel, or employee
- the type of office building used in the data set.

5.6.1   The method of calculating personnel numbers

For reasons explained earlier in this document (see Section 5.2) it is not easy to define a key performance indicator based on occupancy or personnel. The method of calculating personnel has always been identified as a potential risk in producing benchmark data based on per person figures.

For this study it was decided to utilise **total employee numbers** – partly because they are easier to obtain without an in-depth knowledge of employment patterns at a particular office building. Other research teams, such as the Watermark team, have used total full-time equivalent employees, which takes account of part time workers.

However in recent years the encouragement by most companies and by legislation, of flexible working patterns, part time working and remote working means that fewer people than ever will conform to the traditional pattern of office based working for five full days a week.

This means that at any time in an office, a proportion of the employees will not be working (and consuming water) in the office. A simple calculation of employees is not likely to accurately represent the true number of water consumers. The calculation can be made more sophisticated by the calculation of full time equivalent employees, but this can be a time consuming calculation and even then will not take into account more informal arrangements such as remote working. The data used for this research used a simple "number of employees" figure, and the figures are likely to have significantly overestimated the number of water consumers in the offices reducing the overall benchmark water use figure. This may be true of any user of the benchmarks and it may be more appropriate, rather than less, to use a benchmark based on *total employees*.

In order to investigate this possibility, the Environment Agency (who have benchmarked their own offices over a period of years) provided some further data based on their southern regional office in 2003 for comparative purposes. This information, although for one year only, shows that the ratio between total employee and full-time equivalent employee is approximately 1:1.5[13]. This figure would be different for each organisation, but if it were applied to the benchmarks derived by this research, they would be increased to:

- 2.96 m^3/full time employee/year (best or good practice target)
- 5.92 m^3/full time employee/year (typical practice benchmark).

These figures are somewhat higher than the benchmarks proposed by this study, and do illustrate the considerable differences that can result depending on whether total employee or full time equivalent employee figures are used.

5.6.2 The type of building used in the research

The buildings used for this research were those with over 350 employees. The reason for this was that the project team wished to minimise the likelihood of any buildings with shared occupancy. These buildings are substantial in size (probably greater than 3500 m^2) and represent commercial company premises, many of which will be let rather than owner occupied. However this may have had certain implications that would affect the figures:

- due to their corporate nature, the employees are more likely to have available the ability to work flexibly including high technology remote working that may not be available to smaller companies
- job and desk sharing. Employee numbers will be high compared with the total water consumption. In addition, larger offices may have a high proportion of "employees" who spend little or no time in the office, including sales staff
- commercial offices should remain saleable and lettable in the property market, particularly for the important large tenant market. The majority of large offices are refurbished every five to 10 years and this applies in particular to tenanted offices where the core areas such as washrooms are the responsibility of the landlord, who will include this as part of the service charge. It is likely that the sanitaryware will have been replaced in most of these offices within the last ten years and will be up to date and water efficient, particularly with respect to urinal controls and toilet flushing.

7 References

1. http://www.envirowise.gov.uk/

2. Griggs, J.; Shouler, M.; Hall, D. 1996
   Water Conservation related to the built environment. In: Proc. of Water Conservation and Reuse CIWEM Conference 1996

3. Personal communication. Obtained from Building Research Establishment in 2001

4. Personal communication. Obtained from the British Tourist Board in 2002

5. David Bartholomew Associates. June 1999. *Water Use in Hotels.* Unpublished client report for Thames Water

6. Mandix. May 2003
   Water Demand Management in the Hotel Sector through Demonstration Projects. Unpublished Study for Multi-client Group including Environment Agency

7. www.environment-agency.gov.uk/savewater

8. http://www.benchmarkhotel.com/

9. David Bartholomew Associates. August 2000
   Water Use in Offices. Unpublished client report for Thames Water

10. Sustainable Development in Government. Water targets available at:
    http://www.sustainable-development.gov.uk/sdig/improving/targetsc.htm

11. *Waterwise: Good for business and good for the environment.* ND. http://www.environment-agency.gov.uk/commondata/105385/waterwise2001_880654.pdf

12. OGC Buying Solutions. Office Sector Benchmark Report. May 2003 Version 1

13. Environment Agency internal occupancy figures for Southern Regional Offices, 2003. Provided by Environment Agency Autumn 2005

When comparing the current research to previous studies, it is important to note that the majority of previous studies and benchmarks were for governmental or regulatory bodies, that may be anticipated to have slightly older buildings and fewer employees per square metre than are demanded by commercial clients. The only study not to utilise governmental buildings was the David Bartholomew Associates' Thames Water study, which derived the occupancy data from assumptions based on the water consumption per square metre.

5.7 Recommended office benchmarks

Based on the analysis outlined in summary above (and included in more detail on the accompanying data CD) the benchmarks for offices would be proposed as:

Table 5.6   Recommended combined office benchmarks

		Cubic metres per year	Litres per day (assuming 253 days per business year)
Typical use	By employee	4.0 m³/employee/annum	15.8 litres/employee/day
	By area	0.6 m³/m²/annum	2.4 litres/m²/day
Best practice use	By employee	2.0 m³/employee/annum	7.9 litres/employee/day
	By area	0.4 m³/m²/annum	1.6 litres/m²/day
Excessive use	By employee	7.0 m³/employee/annum	27.7 litres/employee/day
	By area	0.8 m³/m²/annum	3.2 litres/m²/day

* Total employee numbers should be used even if full time equivalent employee figures are available.

Note: Figures have been rounded to form more usable benchmark figures.

The project steering group concluded that in order to provide appropriate guidance to the offices sector, it would be useful to provide the typical benchmark only, encouraging facility managers to aim for a figure lower than this. These benchmarks are shaded in the table.